What Time

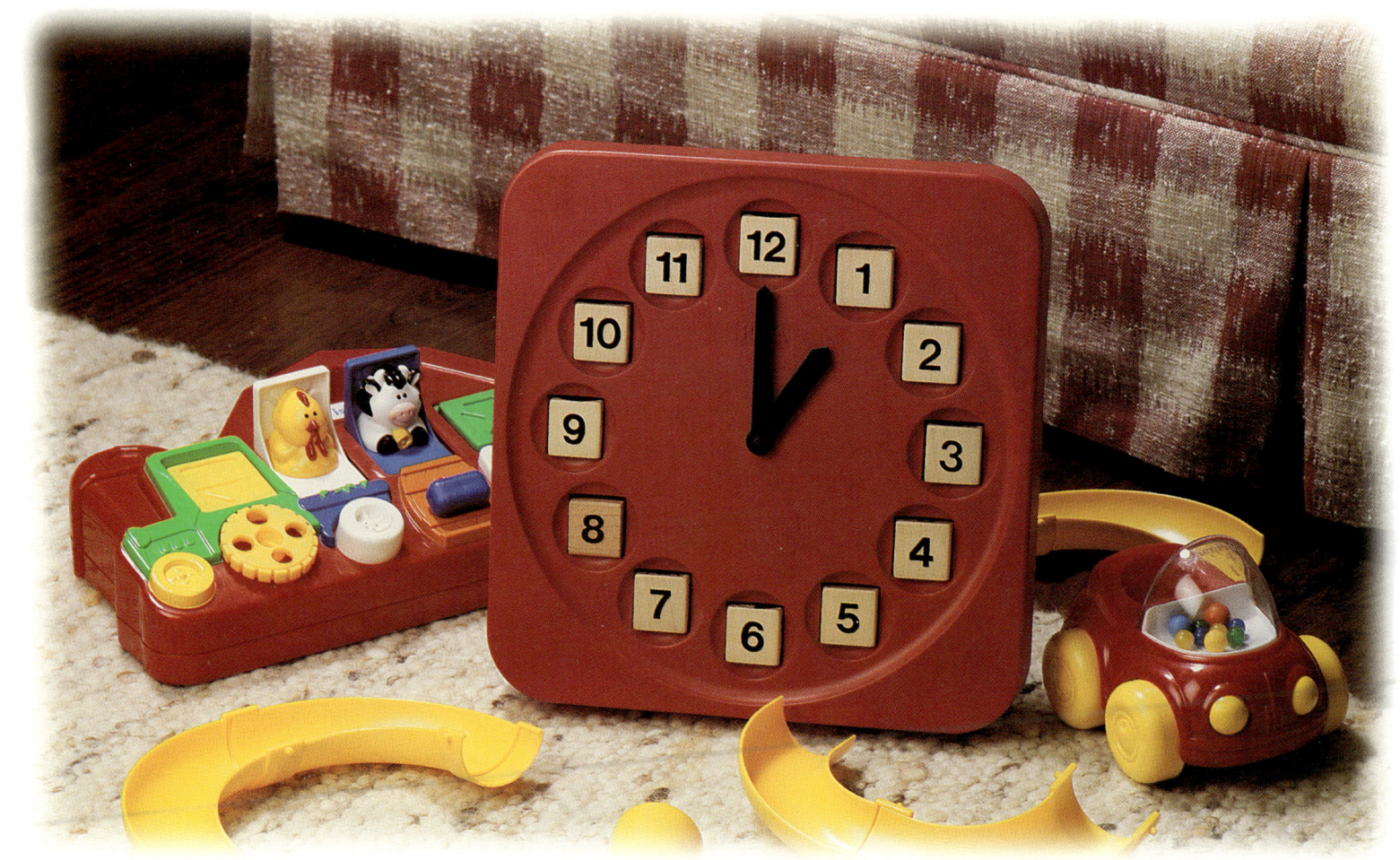

What time is it?

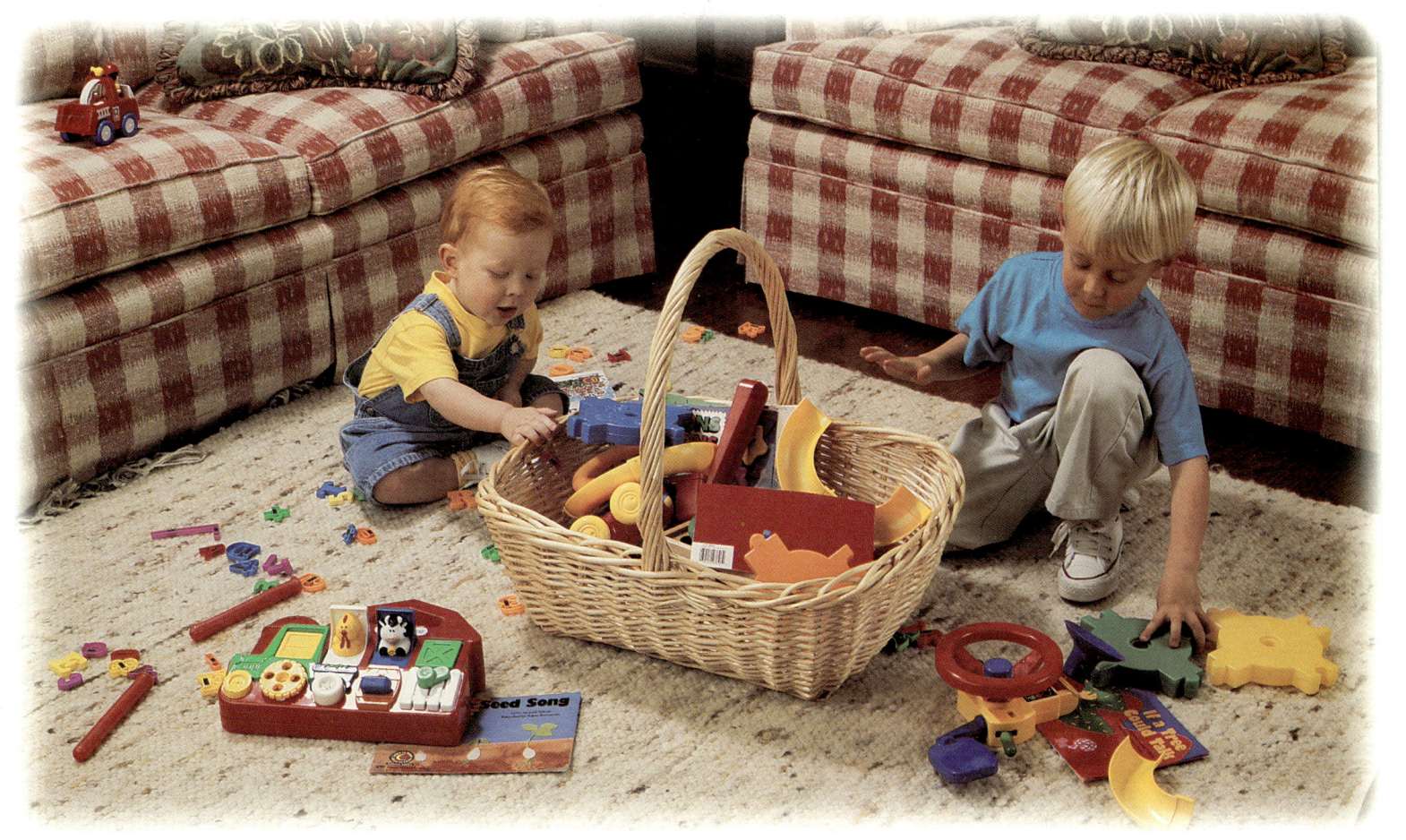

It's 1 o'clock.
It's time to clean up.

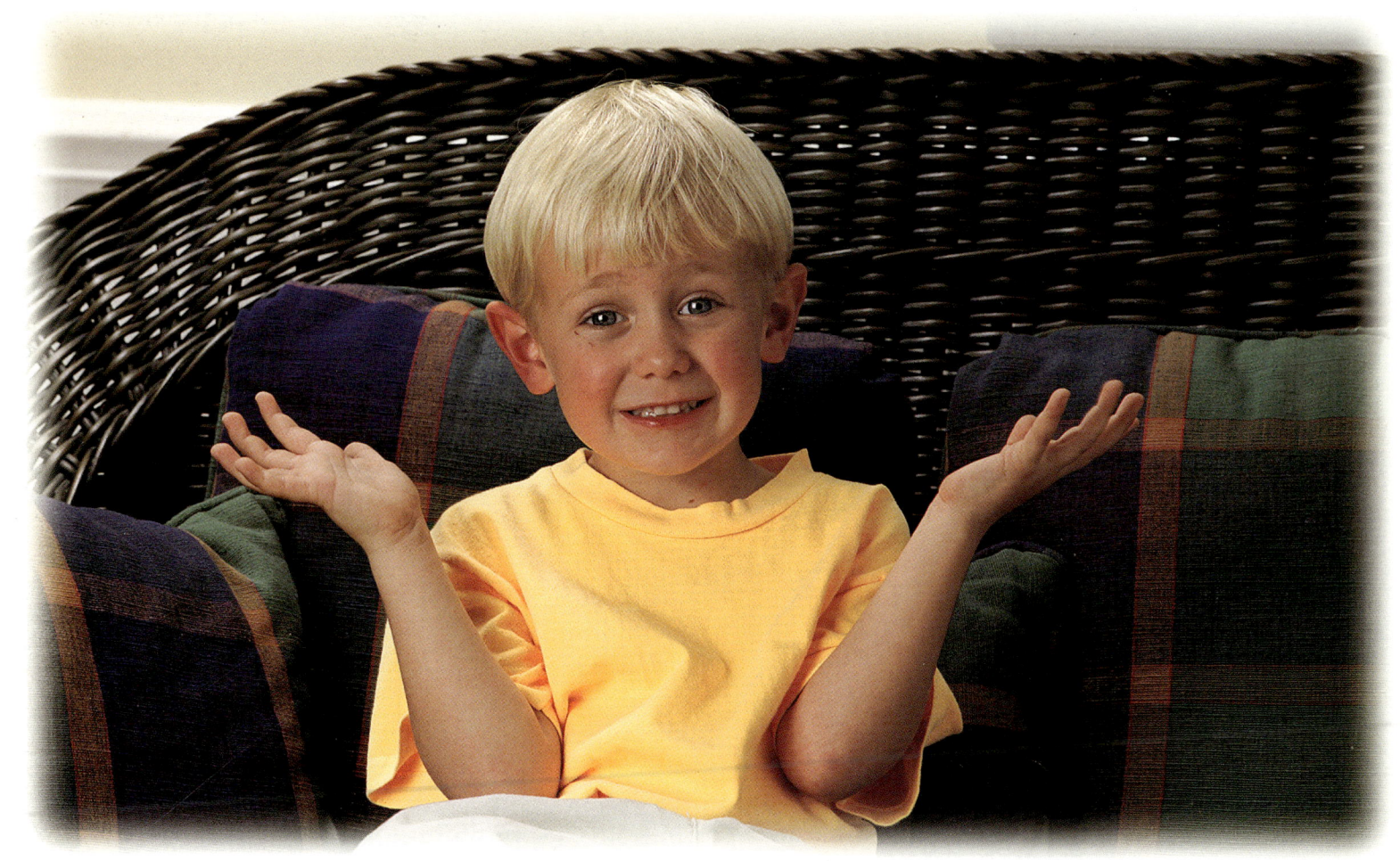

What time is it?

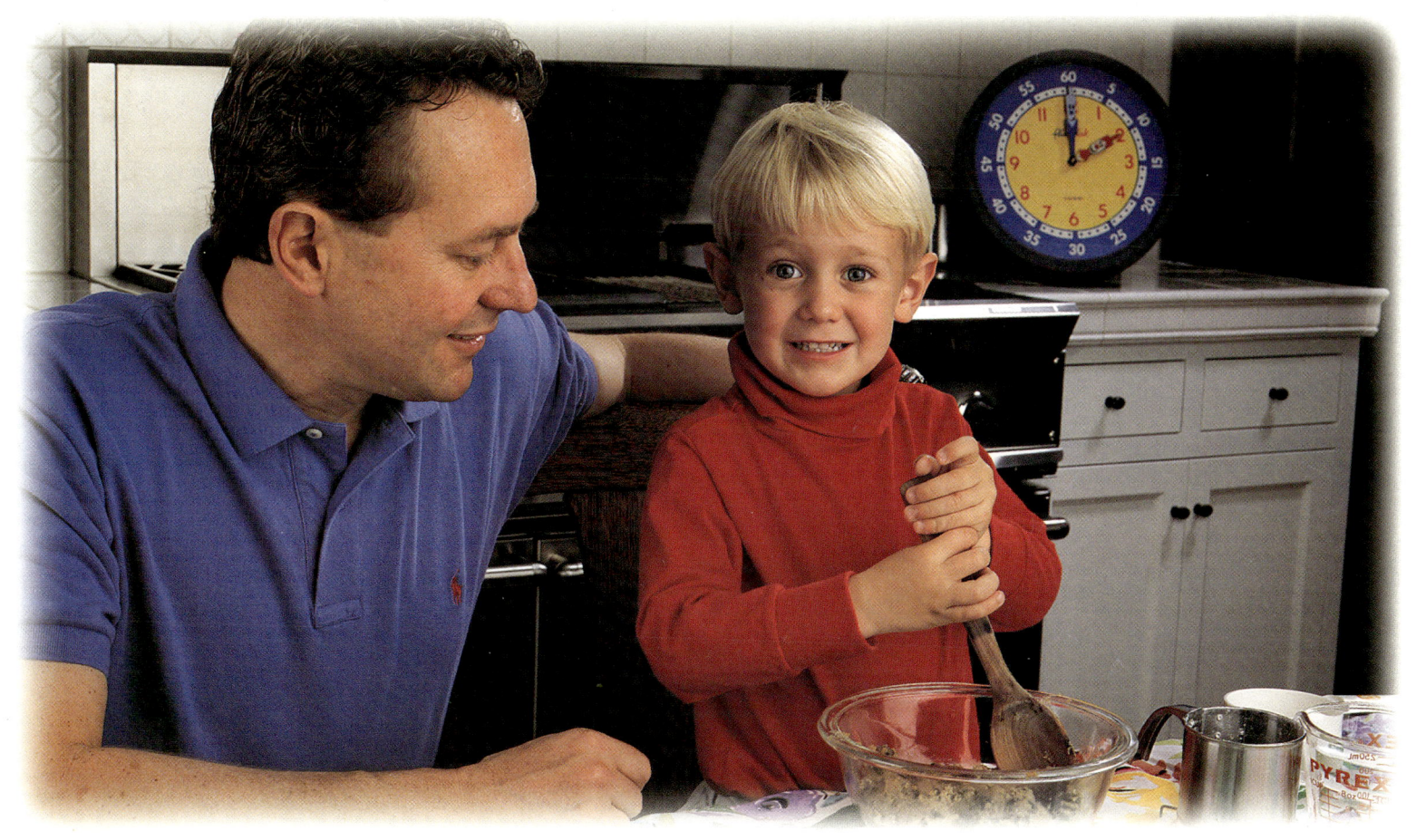

It's 2 o'clock.
It's time to bake cookies.

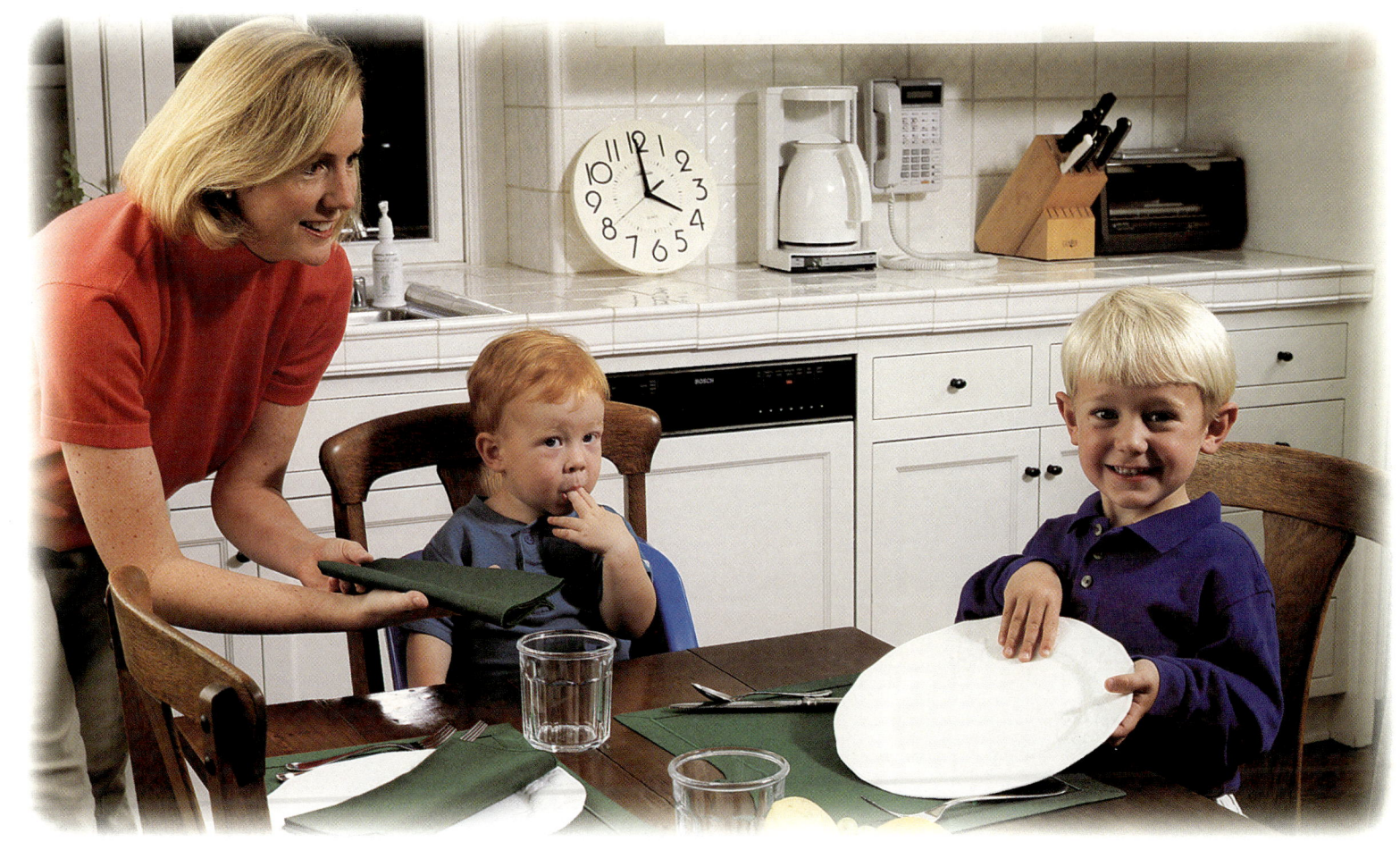

It's 4 o'clock.
It's time to set the table.

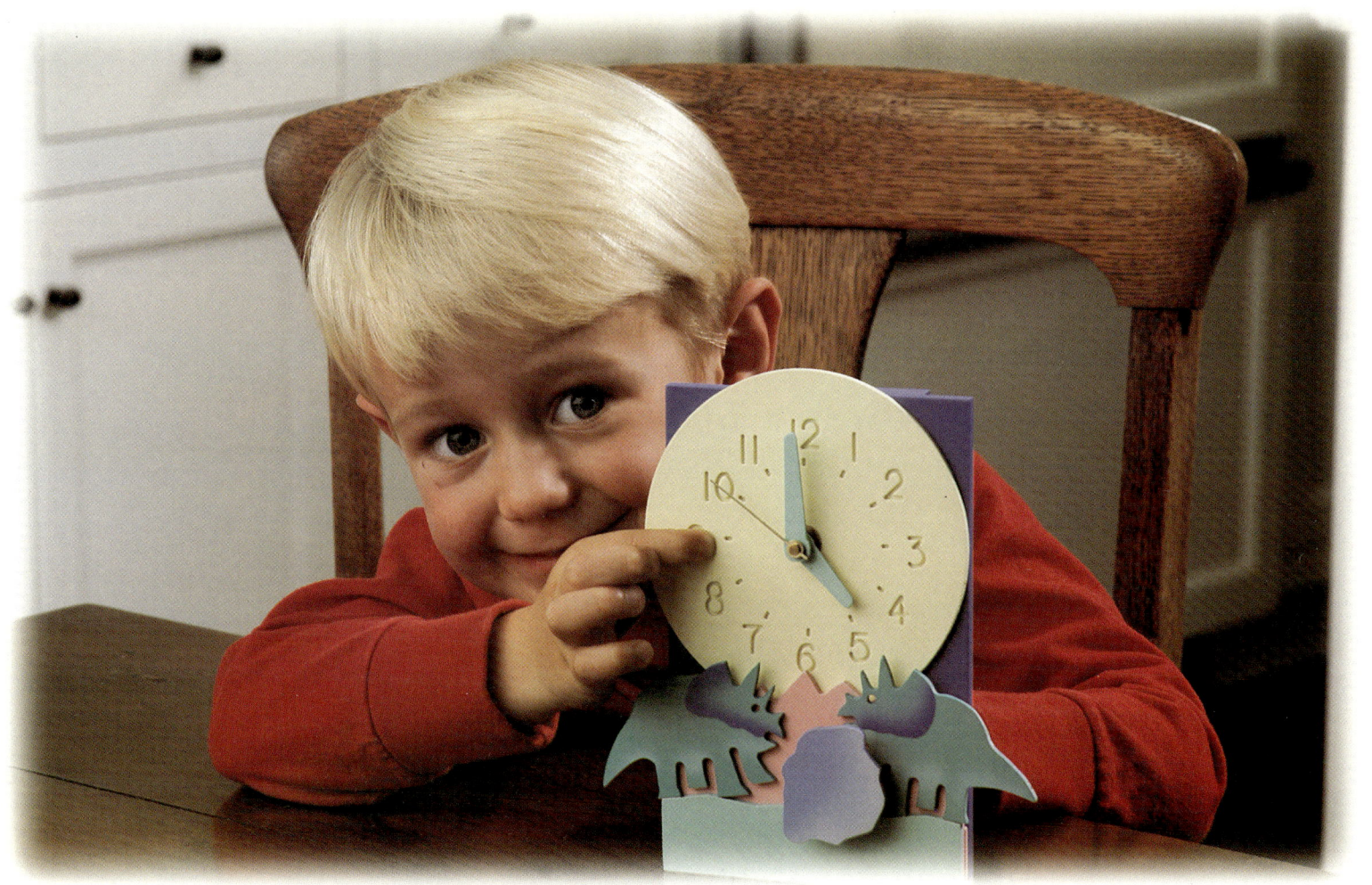

What time is it now?

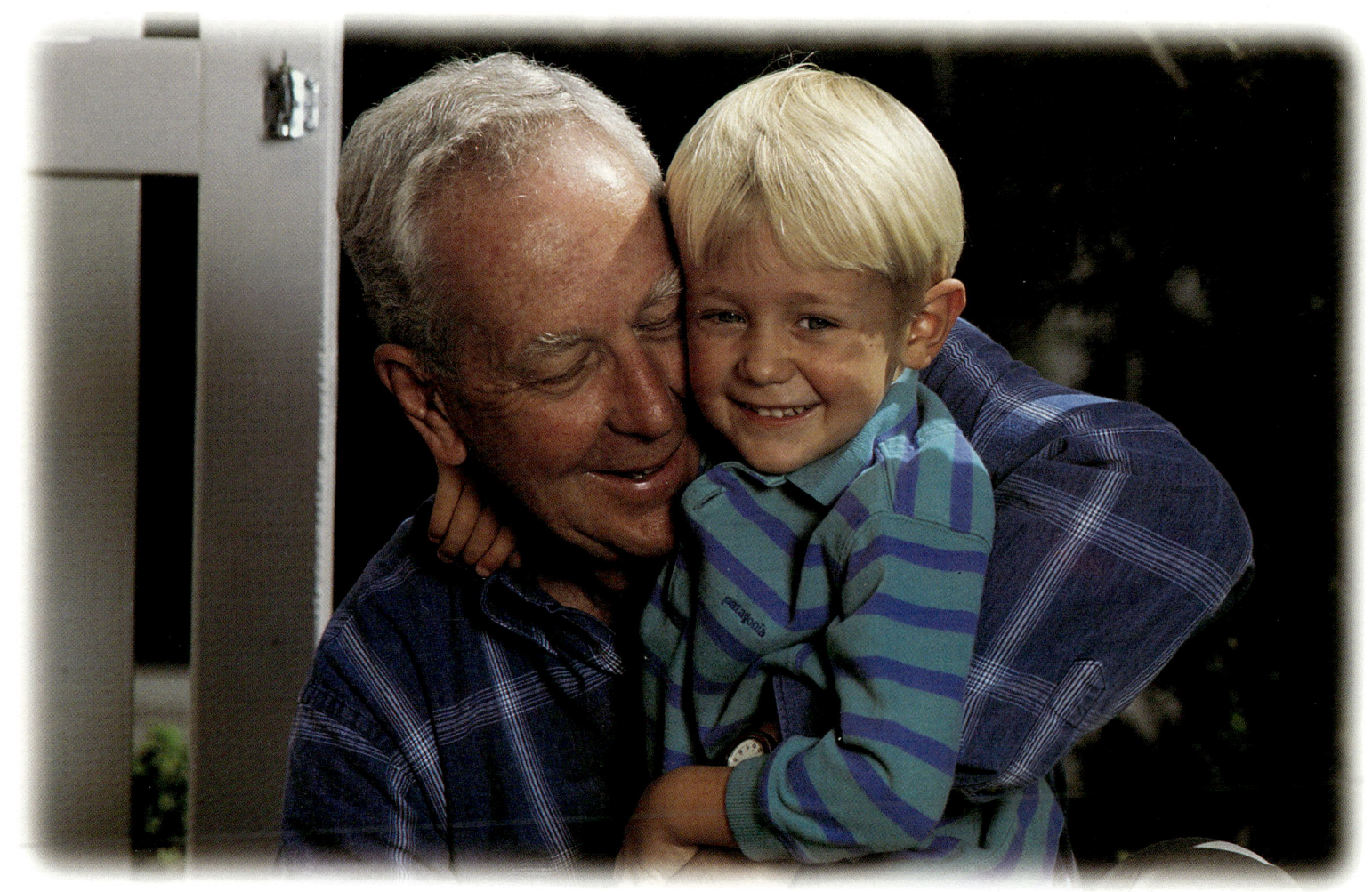

It's time to hug Grandpa!